HAWAI‘I’S White Tern

To Susan,

With aloha,

Susan Scott

SUSAN SCOTT

HAWAI‘I’S White Tern

Manu-o-Kū, an Urban Seabird

A Latitude 20 Book

UNIVERSITY OF HAWAI‘I PRESS
HONOLULU

Printed in China

23 22 21 20 19 18 6 5 4 3 2 1

Library of Congress Cataloging-in-Publication Data

Names: Scott, Susan, author.
Title: Hawaii's white tern : manu-o-Ku, an urban seabird / Susan Scott.
Description: Honolulu : University of Hawaii Press, [2018] | "A latitude 20 book." | Includes bibliographical references.
Identifiers: LCCN 2018022856 | ISBN 9780824878023 (pbk. ; alk. paper)
Subjects: LCSH: Gygis alba—Hawaii.
Classification: LCC QL696.C46 S39 2019 | DDC 598.3/3809969—dc23
LC record available at https://lccn.loc.gov/2018022856

University of Hawai'i Press books are printed on acid-free paper and meet the guidelines for permanence and durability of the Council on Library Resources.

Designed by Mardee Melton

Terns

Outside the hospital window at dawn
I watch in pain a pair of white fairy-terns
weave their intricate ballet over, in,

and around the huge monkeypods and banyans.
They sail exquisitely echoing lines,
the second bird playing tune and variation

on the path of the first, again and again,
until finally they settle into leaves,
but, no, there they go again, soaring.

Later in the morning when I hobble
back to my window they are long gone,
off to the sea for dangerous divings

for the small fish that are their living.
Birds of the sea and of the city,
their lines of flight explain how one thing

can follow another in a lovely way,
with a turn behind a twist,
a rise after a decline.

Parent and offspring, Honolulu.
(RICH DOWNS)

From *A Field Guide to the Wildlife of Suburban O'ahu* by Joseph Stanton. 2006. Courtesy of Joseph Stanton and Time Being Books, St. Louis, Missouri.

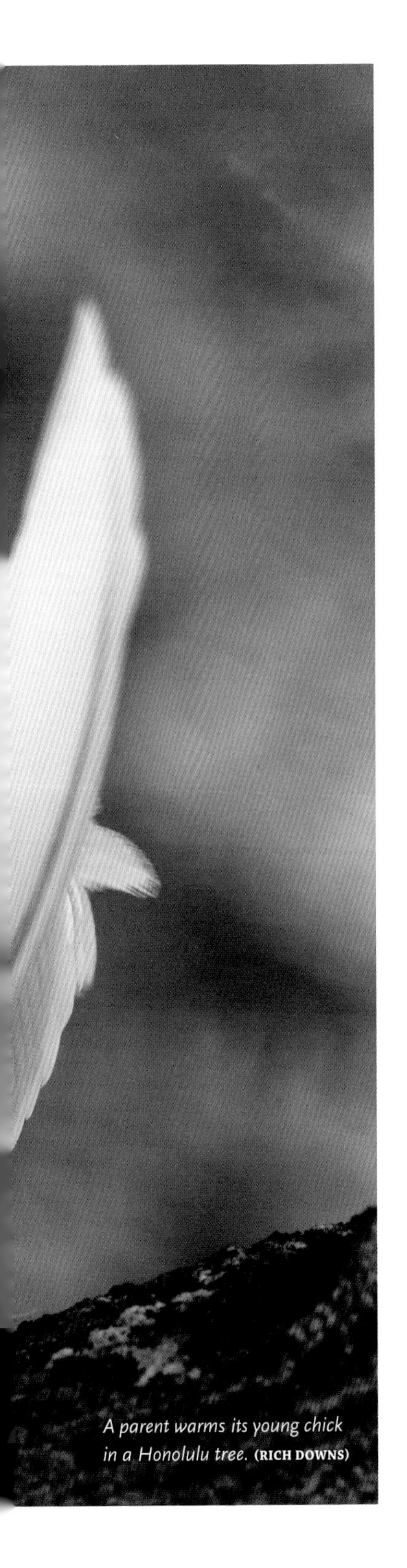

A parent warms its young chick in a Honolulu tree. (RICH DOWNS)

Contents

The two displaced chicks pictured here wait for fish from their human caretakers on Midway Atoll. (DAVID DOW)

Acknowledgments

I fell in love with White Terns during a 1989 visit to Tern Island, where the U.S. Fish and Wildlife Service manager there, the late Ken Niethammer, invited me to accompany him during his monitoring of the island's White Terns. For introducing me to the species and sharing his knowledge and study of the birds he so admired, I will be forever grateful to Ken.

Over the 30 years I've been writing a column for the *Honolulu Star-Advertiser* about marine wildlife, I've received countless letters from readers asking questions about, and sharing their experiences with, Honolulu's White Terns. As the birds expanded their range throughout the city, more people told stories and sent pictures of terns they watched raising chicks in the city's trees. Those readers were the inspiration for this book, and I appreciate all who took the time to write. One reader, Susan Leong, put together a book of the White Tern family that she and her coworkers observed outside their downtown office building in 2014. "I had the privilege of watching a pair of terns raise their chick," Susan wrote, "and wanted to share my photo book with a fellow lover of white terns." The copy she sent me is a treasure.

Dismayed by White Tern eggs rolling off windowsills at Tern Island, French Frigate Shoals Atoll, refuge manager Ken Niethammer chipped hollows in the concrete. (SUSAN SCOTT)

Another reader, Joseph Stanton, professor of art history and American studies at the University of Hawai'i at Mānoa, sent me, unsolicited, his heartwarming poem "Terns," and gave permission to use it. I know all who read the poem join me in thanking Professor Stanton for sharing his lyrical account of how Honolulu's White Terns helped him heal.

My search for information about White Terns happily brought me to Hui Manu-o-Kū, a group of seabird scientists, conservationists, teachers, and other citizens who have come together to "enhance awareness, appreciation, understanding and conservation of Manu-o-Kū" (www.whiteterns.org). I thank Hui members for welcoming me into the group, and for sharing facts from their ongoing work with White Terns. The group's principal science advisor, Dr. Eric VanderWerf of Pacific Rim Conservation, allowed me use of his unpublished data, as did Dr. Alex Wegmann of the Nature Conservancy. I'm grateful to them for helping make the information in this book current and accurate.

Scientific literature is often inaccessible to the public, and when studies are available, they can be difficult to interpret for people without scientific backgrounds. I wrote this book to bridge that gap, and thank Dr. Eric VanderWerf for his help in making the book accurate and understandable. I also thank Professor David Hyrenbach of Hawai'i Pacific University for sharing his extensive knowledge of seabird science and volunteering his time to help me understand wing loading, predictive modeling, and the value of strong coffee.

Rich Downs, Hui Manu-o-Kū's volunteer director, deserves my deepest gratitude for his tireless work in monitoring birds, organizing data, furthering education, and advancing research about Honolulu's White Terns. Rich's commitment to the White Tern cause goes beyond normal volunteer work. After circumstances required him to move from Honolulu to the U.S. mainland, Rich has continued his tern work, flying to O'ahu periodically to collect White Tern data, schedule and run Hui meetings, and perform the nearly endless other duties he has cheerfully assumed. It was a pleasure getting to know Rich, who answered my questions promptly and graciously. Through his sociable tours, sponsored by the Hawai'i Audubon Society, I made new friends and learned much about our city's amazing terns.

Thank you, too, to Keith Swindle, key Hui member and special agent with the U.S. Fish and Wildlife Service Office of Law Enforcement, who for years has been a champion of White Terns and their admirers, both on and off the job. Keith uses kindness and compassion in his work, giving our community a friendly, positive image of wildlife law enforcement. Also thank you to Jason Misaki of Hawai'i's Department of Land and Natural Resources for his support in tree flagging and other research efforts.

Because scientific literature about White Terns is worldwide, and often decades old, gathering research material for this book was sometimes difficult. I thank seabird specialist Dr. Elizabeth Flint, U.S. Fish and Wildlife Service supervisory biologist for Pacific Islands refuges and marine national monuments, for finding and sharing several primary sources that had eluded me.

Hui members, friends, and acquaintances, I thank you for sharing your outstanding photos. All pictures in this book are donations to the White Tern cause.

Rich Downs (center, in white shirt) leading a 2017 White Tern tour on the grounds of ʻIolani Palace. (SUSAN SCOTT)

Photographers provided all photos for free, and readers will see each contributor's name in the captions. The articles published in the Society's peer-reviewed journal, *'Elepaio,* were invaluable in the writing of this book. Finally, as a member of the Hawai'i Audubon Society since 1988, I will donate royalties from this book to that organization, dedicated to protecting native wildlife through science, education, and advocacy in Hawai'i and the Pacific.

It's always hard to start writing a book, especially when information about the subject is continually advancing. For their encouragement to just do it, I am grateful to education specialist Jody Smith and *pueo* (Hawaiian short-eared owl) researcher Dr. Javier Cotín of the University of Hawai'i College of Tropical Agriculture and Human Resources, Department of Natural Resources and Environmental Management, as well as executive editor Pamela Kelley of the University of Hawai'i Press.

Thank you too to seabird expert, and friend, Michelle Hester, executive director of Oikonos Ecosystem Knowledge, for her thoughtful suggestions and moral support, and for beefing up my birding vocabulary and adding pizzazz to my life.

Last but never least, I thank my husband, Craig Thomas, for the time he takes from his full schedule as an emergency physician to hear me out and to offer ideas when I'm stuck. It's great to have a live-in editor, counselor, and friend-who-focuses-on-the-big-picture during the trying times that come with writing any book.

Any errors in this book are mine alone.

City Proclamation

The White Tern (*Gygis alba,* Manu-o-Kū) becomes the official bird of Honolulu

WHEREAS, the white tern occurs throughout the tropical oceans of the world and represents the beauty, simplicity, tranquility, and peacefulness of nature; and

WHEREAS, the white tern is indigenous to, and nests throughout, the Northwestern Hawaiian Islands, but within the main islands occurs solely within the City and County of Honolulu, where it raises its offspring and is a devoted parent; and

WHEREAS, the white tern is known as manu-o-Kū by the Kanaka ō'iwi and was used as a wayfinder and guide to land by the ancient voyagers, and still serves those preserving the ancient Polynesian ways and skills of navigation; and

WHEREAS, manu-o-Kū is recognized and protected by international treaty, federal law, and state law, and is even named in a treaty between Japan and the United States, thereby serving as a symbol of international peace and goodwill; and

"There is something ethereal about the White or Love Tern which seems to remove it from this world." From Bird Flight *by Gordon C. Aymar, 1938.*
(MARTHA BROWN AND BRECK TYLER)

WHEREAS, as a beautiful seabird that calls both the open sea and Honolulu its home, the tern represents an image and quality of life that is uniquely shared by the City and County of Honolulu and serves as an ambassador of our people and our way of life; and

WHEREAS, it is the desire of the City and County of Honolulu to adopt a gift of nature as a symbol of our endeavors to preserve our precious resources and to make our home, our Honolulu, the best place in the world to live, work, and raise our families,

NOW, THEREFORE, I, MUFI HANNEMANN, Mayor of the City and County of Honolulu, do hereby declare the white tern, *Gygis alba,* manu-o-Kū, the official bird of the City and County of Honolulu. Done this 2nd day of April, 2007, in Honolulu, Hawaiʻi. **MUFI HANNEMANN**

Why White Terns Get Their Own Book

Other native seabirds nest on O'ahu and its nearby islands, but the buoyant, graceful White Tern is the only one that views the city's nonnative trees as good places to raise chicks. In laying their eggs in the branches of monkeypod (*Samanea saman*), shower (*Cassia* spp.), kukui (*Aleurites moluccana*), banyan (*Ficus* spp.), mahogany (*Swietenia mahogani*), and other alien trees' natural crooks and unnatural trimming scars, the White Terns of Honolulu have learned to live with people, pets, vehicles, and buildings.

The curious, bold birds sometimes hover over human heads like feathered fairies, and peer into our apartments from lanai railings. Nor do the birds mind us staring back as they court potential mates, tend their precious eggs, and feed fish to their begging babies.

Given these traits, the easily visible White Terns have become a favorite of Honolulu residents who want to know more about the remarkable native species. This book is for them and, of course, for the birds.

Because White Terns breed throughout the tropics and subtropics, facts about the birds come from reports worldwide, including Ascension and St. Helena Islands in the mid-Atlantic Ocean, the Seychelles archipelago in the Indian Ocean, Australia's Norfolk Island in the South Pacific, the islands of Papahānaumokuākea National Marine Monument in Hawai'i's Northwest Chain, and the city of Honolulu.

Due to the diverse environments in which White Terns live, what we know about the birds in one area may not apply to another. This is particularly true of Honolulu's terns, the only population in the world that is thriving, and increasing, in a bustling city. Due to studies targeting Honolulu's White Terns, we know more about the city's population than we do about populations in most places of the world. As researchers and citizen observers report on terns' behaviors, information about O'ahu's birds continues to grow.

A White Tern rests on a resident's railing overlooking the Ala Wai Canal in Waikīkī. (DONNA TAMASESE)

(Above) A White Tern's precarious egg site on Sand Island, Midway Atoll. (DAVID DOW)

(Left) White Terns throughout the world are the same species as Honolulu's White Terns. Pictured is an individual on Cousin Island in the Seychelles archipelago, Indian Ocean. (JAVIER COTÍN)

The book's sections start with known details about Honolulu's White Terns, followed by specifics in the rest of the Hawaiian archipelago, and conclude with data from White Tern researchers in other parts of the world.

Because the White Tern builds no nest whatsoever, but rather lays a single egg on a bare branch or ledge, I use the terms "breeding" and "branch," rather than "nesting" and "nest," to mean laying an egg or tending a chick.

Names

Hawaiian name: Manu-o-Kū

No one knows why ancient Hawaiians called the White Tern Manu-o-Kū. In the Hawaiian language, "manu" means bird, and Kū, with a capital K, is the ancient god of war. The *Hawaiian Dictionary* states that Manu-o-Kū refers to the White Tern and means literally "bird of Kū."

In a 1941 issue of *'Elepaio,* ornithologist George C. Munro speculated that the bird's original Hawaiian name might have been *'ohu,* a word that the *Hawaiian Dictionary* defines as "Mist, fog, vapor, light cloud on a mountain . . ." Munro writes, "This surely describes the beautiful white tern as it flies around one's head."

Under some conditions, the White Tern's wing and tail feathers look semi-transparent. (DAVID DOW)

Common name: White Tern vs. Fairy Tern

A bird's common name is what people in a local community call the bird, but with widespread species, such as the White Tern, common names sometimes differ from one region to the next. In the 1941 article mentioned above, Munro described the bird, *Gygis alba,* as "The White or Love Tern." Some Hawaiʻi residents also called the bird "fairy tern," but that name refers to a different tern species (scientific name *Sternula neresis*) found in New Caledonia, New Zealand, and Australia. In other parts of the world, a common name for the White Tern was "angel tern."

To avoid confusion, the American Ornithological Society (AOS), an organization dedicated to advancing scientific knowledge and conservation of birds, gives birds found in the United States common English names. White Tern is the society's official common name for the species, a designation accepted worldwide.

The AOS recommends that birds' English names be capitalized, a guideline most journals, field guides, and books, including this one, follow.

As a result of their charming appearance, people have called White Terns love terns, fairy terns, and angel terns. The bird's official common English name, worldwide, is White Tern. (SUSAN SCOTT)

Scientific name: *Gygis alba*

As researchers learn more about the evolutionary history of species, scientific names also change to reflect how they are related to one another. In 1786, Swedish naturalist Anders Sparrman named White Terns *Sterna alba.* The first name *Sterna* (the genus) is one taxonomists assigned to about 30 tern species, but later, German zoologist Johann Georg Wagler (1800–1832) decided that White Terns were different enough from other terns that they deserved their own genus. Wagler changed the White Tern's first scientific name to *Gygis,* pronounced J-EYE-jis, from the ancient Greek *guges,* meaning a water bird. The White Tern remains the only bird in the genus. The species name, *alba,* is Latin for white. The AOS's abbreviation for White Tern is WHTE.

Hawai'i hosts two other tern species, the Sooty Tern (*Onychoprion fuscatus*) or *'ewa'ewa,* and the Gray-backed Tern (*Onychoprion lunatus*) or *pākalakala,* and three noddy species called Black Noddy (*Anous minutus*) or *noio,* Brown Noddy (*Anous stolidus*) or *noio kōhā,* and Blue Noddy (*Procelsterna cerulea*). White Terns are more closely related to noddies than to other terns.

Noddies, such as this Black Noddy, get their name from bowing, or nodding, their heads during courtship. Native throughout the tropics and subtropics, including Hawai'i, noddies and White Terns are closely related. (SUSAN SCOTT)

Location

Offshore

White Terns may stay at sea for months, but it's unknown how far they fly from land. When raising a chick, however, White Terns sleep on their island homes at night, go to sea at dawn to fish, and return to the island to feed their chicks, sometimes making several round trips per day.

A White Tern flying with fish shows offshore navigators the course toward land. This parent likely flew to the home branch with fish for its chick, but probably saw something that needed chasing away before the chick could be safely fed. Terns carrying fish must guard against other birds stealing the catch. (DON C. POOLE)

Canoe physician Craig Thomas steers the voyaging canoe Hōkūle'a *in the Indian Ocean with Captain Bruce Blankenfeld behind. Blankenfeld and other traditional navigators use Manu-o-Kū and other seabird sightings to help them find land.* (NĀ'ĀLEHU ANTHONY)

Offshore sailors on traditional voyaging canoes keep a close eye on Manu-o-Kū carrying fish, as well as noddies, or *noio* (which do not carry fish in their beaks). According to master Hawaiian navigator Nainoa Thompson, president of the Polynesian Voyaging Society, "the manu o kū (white tern) go about 120 miles out and noio (brown tern) . . . go about 40 miles out" in their search for fish and squid.

A double-hulled voyaging canoe covers about 120 miles per sailing day in average wind and wave conditions. Thompson writes, "When the manu o kū is fishing, it flutters above the ocean surface, but when the sun starts to go down, it will rise up from the water so it can see farther, and it will head straight back to land. When we see these birds in the day we keep track of them and wait for the sun to get low and watch the bird; the flight path of the bird is the bearing of the island. Then we turn on that bearing, sail as fast as we can, and at sunset we climb the mast to see if we can find the island. And if we can't see it, we heave to [stop the boat] until the morning."

Main Hawaiian Islands

Whether White Terns raised chicks in the main Hawaiian Islands before the arrival of ancient Polynesian voyagers or during the era of ancient Hawaiian settlement is not known. In his 1869 "Synopsis of the Birds of Hawaii," Sandford B. Dole listed White Terns as present in the main Hawaiian Islands, but he did not state that the birds reproduced there.

In his 1941 *'Elepaio* article "Adventures in Bird Study: The White or Love Tern," George C. Munro writes that during a bird collecting expedition in Hawai'i in 1891, Ni'ihau residents "held the bird as sacred and did not care to kill one" for the collection. The account suggests that the birds were known to Hawaiians living on Ni'ihau, but there is no mention of the birds on the other human-inhabited islands.

Nonbreeding birds are hard to count because, without chicks to feed, singles come and go at irregular times and don't stay in one territory. Seabird counters, therefore, usually tally eggs or chicks and report numbers as breeding pairs. The three birds pictured here count as one breeding pair. (CLAUDE MATSUMOTO)

O'ahu

In the 1940s, people noted White Terns flying offshore of O'ahu or passing over the island. In 1961, the first White Terns breeding on O'ahu were noted when an avid birder photographed an adult sitting on an egg in a Kiawe tree near Koko Head. Since then, the White Tern population has steadily increased in Honolulu. Between 1981 and 1988, estimates of birds breeding on O'ahu ranged between 50 and 100 pairs, meaning 100 to 200 individuals.

The tall trees in Kapi'olani Park and Waikīkī's Fort DeRussy Park hosted an early concentration of the birds. By 2002, seabird specialist Dr. Eric VanderWerf estimated that Honolulu hosted approximately 700 adults. About 500 (250 pairs) of those were breeding.

In bustling Waikīkī, on Kalākaua and Kūhiō Avenues, as well as some side streets, White Terns breed in the branches of several kinds of tall trees. Tern parents also raise chicks in trees near ʻIolani Palace and other urban areas hosting tall trees. In this photo, a White Tern parent brings fish to its chick in a tree near the state capitol in downtown Honolulu. (ROBERT WEBER)

In 2004, VanderWerf and Keith Swindle of the U.S. Fish and Wildlife Service Office of Law Enforcement came up with the idea to make the White Tern Honolulu's official bird, thus highlighting the bird's significance as a guide to traditional Polynesian wayfinders, as well as publicizing the bird's presence in the city. The White Tern is the only conspicuous native bird breeding in the trees of Honolulu's parks, yards, and city streets.

In 2005, Swindle shared the idea with Laura Thompson, mother of traditional Polynesian navigator Nainoa Thompson, who suggested the concept to the mayor. The point was well taken, and on April 2, 2007, then-mayor Mufi Hanneman proclaimed the White Tern the official bird of Honolulu.

White Tern numbers in Honolulu continue to rise. In 2016, Dr. VanderWerf and volunteer Rich Downs counted approximately 2,300 White Terns in the trees of Honolulu, from Niu Valley in the east to Hickam Air Force Base in the west. These researchers found that the increase in the number of terns on the island since 2002 was not due to the birds significantly expanding their range, but rather that more birds are breeding in areas already occupied.

Several theories exist about the reasons for Honolulu's population increase, one being that waters off Honolulu may be rich in fish and squid, enabling the birds to successfully find food close to home. Another factor may be that fewer introduced tern predators—rats, cats, and barn owls—inhabit the urban area.

So far, Oʻahu is the only main Hawaiian Island chosen by White Terns to raise chicks, and their range on the island is narrow. For reasons unknown, pairs prefer (so far) the south side of the island, from Niu Valley to Sand Island, even though the tall trees the birds favor grow throughout Oʻahu. The White Terns' range, however, has been slowly widening over decades. As chicks mature and return to the island to find their own breeding sites, the birds may continue to spread out.

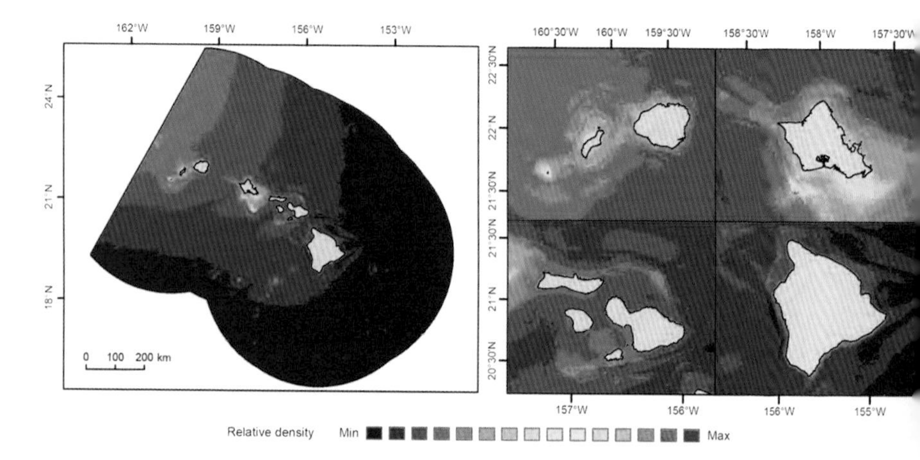

These computer models predict, by color, where White Terns likely fly off the main Hawaiian Islands. Orange indicates the most likely areas, dark blue the least. Models were based on shipboard sightings between 1989 and 2012, mostly between August and November, at that time low breeding months for White Terns. During approximately 12,280 miles of cruising, researchers saw only 86 White Terns. Factors affecting the bird's distribution were distance from shore, water depth, latitude, and longitude. (NOAA TECHNICAL MEMORANDUM NOS NCCOS 214)

Northwest Hawaiian Islands

White Terns are abundant in the Northwest Hawaiian Islands, established as the Papahānaumokuākea Marine National Monument in 2006. The 1,200-mile chain stretches from Nihoa Island at 23.1 degrees north latitude to Kure Atoll at 28.4 degrees north latitude. Kure Atoll is the northernmost White Tern breeding spot in the world. The approximate number of White Terns inhabiting the Papahānaumokuākea Marine National Monument's nine islands is 80,700.

Since 1903, Midway Atoll's 1.8-square-mile (486 ha) Sand Island has hosted various private and military

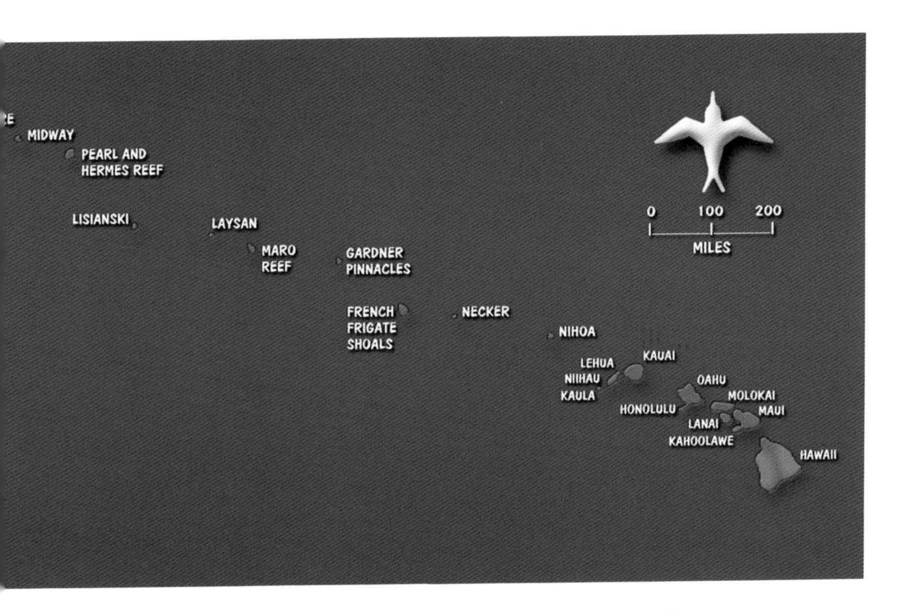

In the Hawaiian Island Chain, White Terns breed on O'ahu, Nihoa, Necker, French Frigate Shoals, Gardner Pinnacles, Laysan, Lisianski, Pearl and Hermes Reef, Midway, and Kure. (USFWS)

operations with island workers and residents planting, among other species, countless ironwood (*Casuarina litorea*) trees. The introduction of ironwood trees and the construction of buildings, both of which provide egg sites for White Terns, caused Midway's population to increase from a few birds to hundreds and eventually to thousands. In a 2017 environmental assessment report concerning a mouse eradication project at Midway Atoll, the U.S. Fish and Wildlife Service estimates that the number of Manu-o-Kū that live on Midway's Sand Island is 50,000.

After the Battle of Midway in 1942, the U.S. Navy dredged coral from the reefs surrounding Tern Island in

A 2015 project to remove lead contamination on Midway Atoll's Sand Island required cutting down several ironwood trees containing about a dozen White Tern chicks. Workers fed the relocated chicks lagoon-caught minnows several times a day. All the chicks fledged, but as they do when raised by tern parents, the chicks occasionally returned home for food. If a worker held up a fish, fledglings would often take it in flight. This photo shows one of those chicks on volunteer tern feeder David Dow's hat, no doubt hoping for a free meal. **(DAVID DOW, SELFIE)**

The two displaced chicks pictured here wait for fish from their human caretakers on Midway Atoll. **(DAVID DOW)**

French Frigate Shoals Atoll, building a naval air station and increasing the island's size from 11 acres (4.5 ha) to 57 acres (23 ha). On the enlarged island, workers planted magnolia (*Magnolia* spp.), beach heliotrope (*Tournefortia argentea*), and ironwood trees (*Casuarina litorea*).

It took 23 years for White Terns to breed on the expanded island. In 1965, a female laid the first White Tern egg on the branch of a Tern Island ironwood tree. Before that, White Terns had been raising chicks only in the cliffs of La Perouse Pinnacle, a one-acre rock rising 130 feet above sea level six miles south-southeast of Tern Island.

Worldwide

White terns range throughout the world's tropical and subtropical regions. The tropics are defined as areas of Earth that lie between 23.5 degrees north latitude, called the Tropic of Cancer, and 23.5 degrees south latitude, called the Tropic of Capricorn. As a climate classification, the subtropics have no distinct line. Although wind, ocean currents, and land masses can cause unevenness, the areas between the approximate latitudes of 23.5 and 35 degrees north and south can be considered rough boundaries of the subtropics.

Some Hawaiian Islands lie in the tropics, some in the subtropics. The southernmost point, Ka Lae (also called South Point) on Hawai'i Island, is 18.9 degrees north latitude. Honolulu International airport is 21.3 degrees north latitude. Kure Atoll is the Hawaiian Islands' northernmost point at 28.4 degrees north latitude.

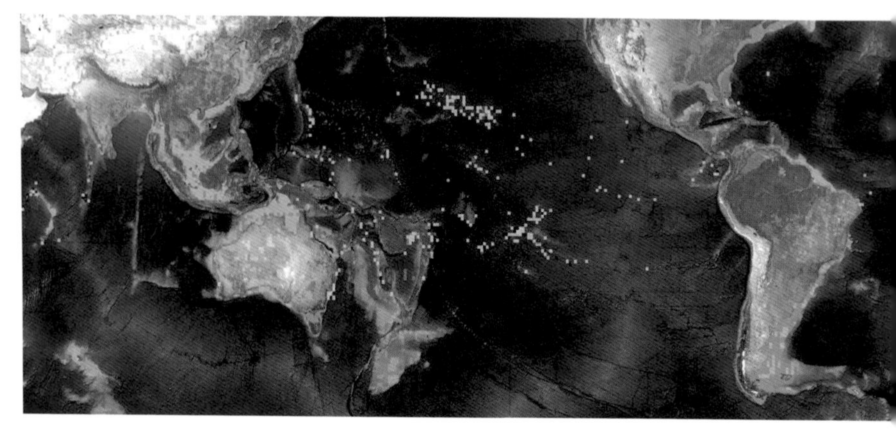

The purple and lavender squares on this map show sightings of White Terns year-round from 1900 to 2017. The colors' intensity indicates the number of reported sightings, with light lavender being fewer birds, and purple being the most. From the Cornell Lab of Ornithology, Birds of North America, White Tern, Distribution, Migration and Habitat, online at https://birdsna.org/Species-Account/bna/species/whiter/distribution. Map generated from eBird observations. GOOGLE MAP DATA, 2018.

White Terns are native in the south Atlantic, Indian, and western and central Pacific oceans, breeding on islands that offer predator-safe sites. The birds have been occasional visitors to Mexico's Pacific islands, and in 2014, researchers noted the first White Tern egg and chick on Socorro Island, 19 degrees north latitude, about 360 miles off Mexico's west coast.

Although it's impossible to count birds that have such a wide global range and breed on remote islands, a rough estimate of the world population is 200,000 individuals. The species does not appear, at this time, to be threatened with extinction.

Description

Size

From the tip of the bill to the tip of the tail, adult White Terns range from 10.8 to 13.0 inches (27.5–33 cm) long. The bird's wingspan ranges from 27.6 to 34.3 inches (70–87 cm). To picture the bird's size, imagine a creature that is a one-foot ruler long and nearly a yardstick wide.

A Laysan Albatross chick and White Tern adult size one another up at Midway Atoll. (DAVID DOW)

The White Tern's sharp-pointed bills are blue-black with a dark blue base. (ALEX WEGMANN)

Hawaiʻi's White Terns have tapered bills averaging 1.5 inches (3.7 cm) long. Adults weigh from 2.7 to 5.5 ounces (77–157 g). For comparison, an average deck of cards weighs 2.7 ounces and a billiard ball weighs 5.5 ounces.

Colors and Feathers

An adult White Tern's skin is black, a color obscured by the bird's all-white feathers. Its black eyes seem larger than they are due to a surrounding circle of black feathers.

A White Tern's legs are slate blue with cream-colored webs between the toes.

White Terns have no seasonal change in feather colors. Males and females look identical.

Feathers are waterproof, a state maintained by the bird regularly collecting oil from a gland at the base of its tail onto its beak and spreading the oil through its feathers. In the process, called preening, the bird straightens

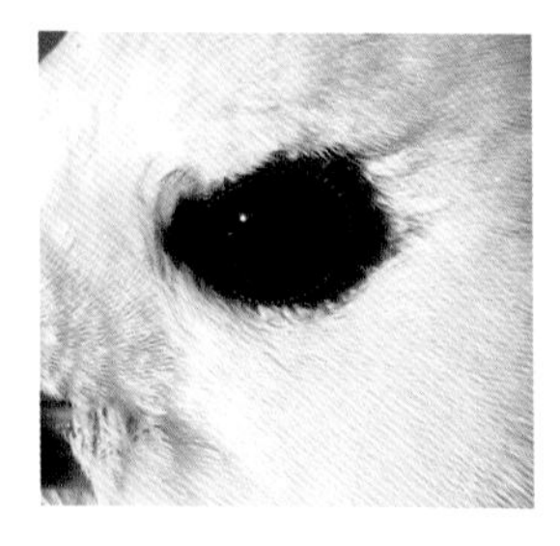

Black feathers and black skin around the eyes make them appear larger than they are. (ALEX WEGMANN)

White Terns' feet are only partially webbed. (SUSAN SCOTT)

Male and female White Terns look the same. (SUSAN SCOTT)

This preening tern is using its beak to align and waterproof its feathers. (JAVIER COTÍN)

Researchers categorize White Terns as flappers, as opposed to gliders such as albatrosses and booby birds. The White Tern's bouncy flight patterns are distinct from those of cattle egrets and pigeons, Honolulu's other white birds. (DAVID DOW)

feathers that have become separated during bouts of fishing and flying.

Among seabirds, Manu-o-Kū have exceptionally efficient molting patterns that enable the birds to maintain their agile flight, and thus raise chicks, year-round. Each large flight feather grows in completely before the next one is shed, ensuring that the slot left by the shed feather is partly covered by adjacent feathers. By having two or three widely spaced feathers growing in the wing simultaneously, the wing's shape remains uniform at all times.

Just before egg laying, both parents lose feathers on their bellies, creating bare skin patches that help transfer the adults' body heat to the egg, and later to the newly hatched chick.

Flight

It's easy to identify a White Tern's graceful, nearly floating flight. From below, the wing feathers of a White Tern hovering overhead can look semitransparent, making the bird resemble a mythical fairy—hence the former name "fairy tern." The birds are capable of making dramatic changes in direction and speed.

Cattle egrets and pigeons, two other white birds (some pigeons have all-white feathers), also fly throughout Honolulu, but these nonnative birds' body shapes and flight patterns are unlike those of White Terns. The fluttering flight of White Terns is distinct and, once you know it, unmistakable.

White Terns have the ability to hover in one place as they scrutinize something that has caught their attention, in this case the photographer. (DAVID DOW)

Age

Researchers determine a bird's age by placing a numbered metal band on a chick's leg using special pliers. The federal government's North American Bird Banding Program, based in Patuxent, Maryland, provides 30-some band sizes, and keeps track of the number, species, place, and date of each banding. When a banded bird is injured or dies, and a person reports the bird's leg band number, workers record the information. Reporting a band recovery (online at https://www.pwrc.usgs.gov/bbl) helps scientists gather facts about the birds.

To band a bird, workers approach a chick when its legs are adult-sized but the youngster is not yet able to fly. Because seabirds are long-lived, and because information is lost if the bird dies somewhere unobserved, longevity studies can take decades. Currently, band recoveries show that White Terns usually live 16 to 18 years, but there are exceptions. On O'ahu, the oldest known White Tern is 37 years old. On Midway Atoll, one bird lived 36 years and another lived 30-plus years (the second bird's age at banding was unknown).

Banding Honolulu's White Terns often requires professional tree climbing. Here, arborist Lake Gibby reaches for a tern chick, upper right, to bring down for banding. (RICH DOWNS)

Workers place a numbered band on a tern chick's leg. (ALEX WEGMANN)

The metal leg band on this parent incubating an egg in Honolulu shows from below. Each time the parent returned from fishing, the band had turned, showing a different part of the number. A photo series eventually revealed the whole number and the bird was identified. It had been banded as a chick near Diamond Head on August 23, 1981, making it, as of this writing, 37 years old. (RICH DOWNS)

The large chick, right, is likely chirping as it eagerly awaits the fish in the parent's beak. (ROBERT WEBER)

Couples occasionally make clicking sounds while swaying on a branch with heads down and eyes closed. The meaning of the sound and motion is known only to the birds. The bird on the right, with bill down and eyes closed, is either rocking or napping. (DAVID DOW)

Sounds

White Tern calls have been described in publications as "grich-grich-grich" and "eenk-eenk-eenk." Each bird watcher has his or her own description of the birds' sounds. Sometimes, a paired, perched couple may point their bills downward, close their eyes, rock back and forth, and make a soft clucking noise in rhythm with the rocking. No one knows what this communication means to the birds. When one or both parents are in sight, especially when carrying fish or squid, a chick often makes soft chirping sounds as a request for food.

Diet

White Terns belong to a seabird group known as "tuna birds" that includes noddies, shearwaters, and other terns. The name comes from the birds' foraging technique. When hunting, tunas, dolphinfish (*mahimahi*), other predatory fish, and dolphins drive small fish and squid to the water's surface, making them available to seabirds hunting above. Several bird species converge to take advantage of such feeding opportunities, creating dense, noisy groups. Because it's a clue to the location of large fish, anglers look for these feeding flocks, calling them "bird piles." The splashes of predators fishing from both above and below, and of prey fighting for their lives, are loud and visible.

Most bird piles in Hawai'i waters are associated with skipjack tuna, known as *aku* in Hawaiian.

White Terns eat mostly fish and some squid. About a quarter of White Terns' fish catch is juvenile flying fish and another quarter juvenile goatfish. The rest are mixed species of fish and flying squid. White Terns sometimes hover over jacks (*ulua*) near shore because hunting jacks chase small fish toward the surface.

In a 1983 study in the Northwest Hawaiian Islands, researchers examined the catches of 241 White Terns. In addition to the above, the birds caught herring, needlefish, halfbeaks, and dozens of other species. It appears that any type of creature the bird can catch and carry crosswise in its beak, including shrimp, is fair game.

One charming trait of White Terns is their ability to hold fish crosswise in their beaks and still catch more. Ridges inside the upper bill help explain the remarkable skill of keeping hold of a catch while continuing to fish.

The prey in this so-called "bird pile" are in a lose-lose situation. To escape predators from below, fish and squid flee to the surface, where seabirds pick them off. (SUSAN SCOTT)

A parent with a two-fish (species unknown) meal waits for its fledged chick to return. (RICH DOWNS)

A rough, flexible tongue (lower, pink) holds fish against the upper beak, securing caught prey while the bird catches more. (ADAM LEE)

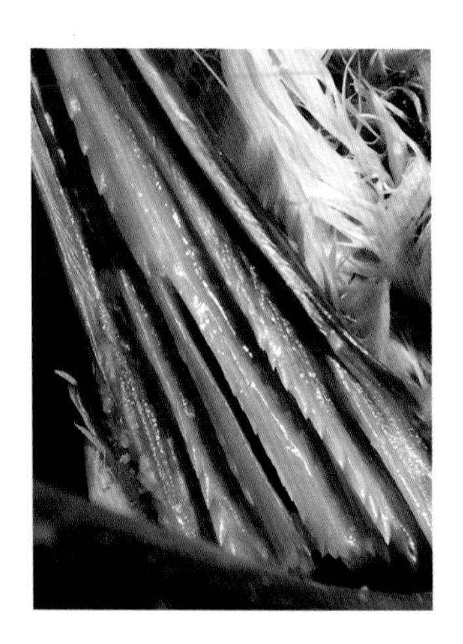

Toothlike structures, shown here, line a White Tern's upper beak. (SUSAN SCOTT)

Puffins (*Fratercula* spp.), seabirds native to cold climates, also do this. The bird's raspy tongue holds fish against toothlike ridges on the inside of the upper beak while the bird carries on fishing. The agile tongue helps parent birds feed their chicks one fish at a time.

The average size of White Terns' fish catch is just under 2 inches (46 mm), ranging from a tiny 0.5-inch-long (12 mm) snake mackerel to a whopping 8-inch-long (200 mm) needlefish. For the most part, White Terns will eat anything they can catch, swallow, or carry.

This parent waits for its chick to swallow a squid before offering a fish. (RICH DOWNS)

Even when parents have their beaks full of fish, they feed the chick only one at a time. (RICH DOWNS)

The challenge for this chick is figuring out how to swallow a needlefish as long as its own body. A parent sometimes helps a chick with such an oversized meal by holding up the fish's tail. (DAVID DOW)

The gossamer feathers of the White Tern may make the bird less visible to prey below, especially at dawn and dusk. (DAVID DOW)

Because White Terns carry their intact catch sideways in their beaks, researchers can identify fish species that are otherwise hard to find. Late author and former Waikīkī Aquarium director (from 1940 to 1972) Spencer Tinker wrote in *Fishes of Hawaii,* "[Gregory's fish] is known but from a single specimen about two inches in length from Laysan Island which was 'brought to the nest of a white tern' on May 12, 1923. This is an example of the extreme depravity to which scientists will descend to obtain a new species, namely, taking food from a little bird."

While fishing, a White Tern scans the ocean surface, and, when it sees a potential prey, hovers there with its bill pointed down. At the right moment, the bird folds its wings, dropping to snatch the fish or squid. In another technique called air-dipping, the bird catches prey in midair as it leaps from fish and mammal

Seabirds take advantage of the energy in wind. Data from 114 cruises between 1976 and 2006 showed that as wind speed increases, terns, as a group, flap more. White Terns can fly without wind. (DAVID DOW)

predators below. White Terns drop to the surface and sometimes plunge a bit below, but rarely, if ever, sit on the water or dive. Although the tern's feet are webbed, they're too small and weak to propel the bird along the surface of the ocean or underwater.

Often, White Terns leave their trees at sunrise and come home at sunset, but sometimes parents make two or more fishing trips in one day. White Terns' translucent feathers likely make them less visible to fish in the low light of early morning and late evening. Food samples from the birds returning from fishing at these times showed a preponderance of squid, lantern fishes, and light fishes, species that spend daylight hours in the deep and come to the surface at night.

White Terns drink neither fresh water nor seawater. Their fish and squid diet provide all the water the birds need.

Breeding

Pairing

White Terns tend to stay with the same mate for several breeding seasons (perhaps for life) and remain around the same chick-raising site year after year. No one knows whether paired White Terns stay together while fishing, return to Honolulu paired, or return as singles and proceed to look for a mate. If two adult White Terns are next to one another in a tree, they are either courting or are a mated pair.

No one knows at what age Honolulu's White Terns begin to breed. In the Northwest Hawaiian Islands, White Terns start breeding at five years old.

Presenting a fish is a courtship gesture between adults, perhaps showing a potential mate the bird's fishing skill. Courting can be brief or go on for weeks. (CLAUDE MATSUMOTO)

A tern preens its mate. (DON C. POOLE)

While courting and/or reestablishing a bond, a White Tern sometimes presents a fish to its mate. A mated pair will also preen the head feathers of the other, since a bird can't reach those feathers with its own beak.

Egg site

A White Tern egg in a natural triangle of branches. (SUSAN SCOTT)

In Honolulu, most White Terns lay their eggs in trees, either in natural crooks of branches, wedged against flaking bits of bark, or in scar cups that form on trunks after trimming. City birds lay eggs and roost in a wide variety of trees, using,

An angry White Tern scolds an intruder. (RICH DOWNS)

in this order, monkeypod (*Samanea saman*), shower (*Cassia* spp.), kukui (*Aleurites moluccana*), Chinese banyan (*Ficus microcarpa*), mahogany (*Swietenia mahogani*), and Indian banyan (*Ficus benghalensis*). Most tern trees are large, with trunks about 20 inches or more in diameter. Mature Indian banyans grow aerial roots that form numerous trunks.

Neither kukui trees nor mahogany trees are common in Honolulu, yet terns seem to seek them out, probably because those trees offer prime egg-balancing opportunities. The sharp, natural bends in kukui branches create hollows that hold eggs. Mahogany bark grows in flakes attached to the tree on one end, jutting out on the other. Such bark barriers can keep tern eggs from rolling off.

City terns will share trees, but sometimes fight over egg-laying sites. When trying to take over a prime egg-laying spot, White Terns have been seen inflicting fatal injuries on members of their own species. Between 1983 and 1985, an observer saw a pair of White Terns in Kapiʻolani Park attack three chicks and an incubating adult. One chick suffered a broken wing, and another fell to the ground.

Although Honolulu birds so far choose to lay their eggs in trees, they aren't above setting up housekeeping on human-built structures. In 2017, a pair of White Terns selected the cracked third-floor lanai railing of the Hawaiʻi State Art Museum on Hotel Street to raise their chick, delaying an eight-million-dollar renovation.

Pictured is a parent covering its newly hatched chick on the Hawaiʻi State Art Museum's third-floor railing. To protect the tern family, State of Hawaiʻi officials delayed a costly upgrade until the chick flew away. (CATHY YOUNG)

Seeing only two terns in a tree is common, but in areas with high tern numbers, breeders and nonbreeders sometimes share an ideal tree. Researchers in 2017 counted nine breeding pairs in one Honolulu banyan tree as well as several nonbreeding terns.

Both mates are sometimes seen working together in choosing a place to lay their egg. Prospective parents sometimes pace back and forth over a spot using small, fast steps while wagging their tails over the site, as if testing for stability. This can go on for weeks before the female lays her egg. Other females lay the egg quickly without seeming to inspect the site at all.

In the Northwest Hawaiian Islands, White Terns lay eggs on low branches and coral rock. The birds also view window ledges, boat trailers, concrete posts, circular

This chick grew up on a Boston Whaler trailer (the boat's hull is visible above the chick's head), preventing Tern Island workers from using the boat until the chick fledged. (SUSAN SCOTT)

water faucets, and other manmade structures as suitable egg-laying sites.

White Terns build no nest whatsoever, laying and incubating the egg in any shallow depression without so much as a twig or feather for cushioning or stability. This may help the birds avoid nest parasites, or it may be an adaptation to cliff nesting where vegetation isn't available.

Egg laying

The female lays a single smooth, mottled egg of tan, brown, and white. On Tern Island, this usually takes place in the morning. The average egg weighs about 0.75 ounce (21.6 g), close to the weight of an AA battery, and measures 1.6 inches (41 mm) long, a tad shorter than a large paper clip. The egg is about one-fifth the average bird's weight.

Within a single breeding season, a female can lay as many as six eggs, but she lays only one for each reproductive attempt. If the egg falls, or a predator eats it, the female will lay another. No data exist as to how long it takes for Honolulu females to replace a lost egg. On Tern Island, the average is 21 days. The range is large, though, with some females laying another egg in 10 days and others taking up to 57 days to re-lay.

Unlike most seabirds, White Terns breed year-round. The pair raise only one chick at a time. Some couples raise three chicks in one year, others raise two chicks, and some just one. One Oʻahu pair was seen incubating an egg, while at the same time feeding a recently fledged chick that still showed wisps of down.

Occasionally, a pair will take a year off from parenting. Between breeding times, some White Terns remain near their home island, and some fly to sea to parts unknown. Even though the birds breed all year, they have peak months that vary from year to year and region to region. In the past, Honolulu birds laid the most eggs in March. In 2016 and 2017, however, a second peak in the number of eggs laid occurred in October. Time will tell if the city's terns will continue to have two bursts of egg laying per year.

In the Northwest Hawaiian Islands, a single peak breeding time usually occurs in April and May.

Incubation

Either the male or female begins incubating their single egg immediately after laying, warming the egg with the parent's bare-skin belly patch.

(Left) This remarkable photo shows a female in the process of laying an egg. (RICH DOWNS)

(Right) This egg in a Waikīkī tree gets incubated one side at a time. Parents turn their egg often, warming it through and through. (SUSAN SCOTT)

Parents take turns sitting on their egg, the timing likely dependent on the fishing success of the partner. Average shifts are between 24 and 48 hours. (SUSAN SCOTT)

The parent has a precarious foothold as it incubates its egg. (SUSAN SCOTT)

Males and females share the sitting equally, usually staying 24 to 48 hours. Shift lengths vary according to couple, however, with some changing after six hours and others brooding as long as 72 hours. The timing presumably depends on the absent parent's fishing success. Because the egg is often balanced precariously, parents slowly raise and lower their bodies from the egg.

After the first day of continual sitting, parents may leave the egg for several hours at a time. An unrelieved parent may leave its egg up to three hours, presumably to find food. One report showed that parents left eggs unattended for a few hours to two days after laying. The eggs still hatched and produced healthy chicks.

If an egg fails to hatch, inexperienced parents may continue sitting on it for up to 180 days. Experienced parents, though, abandon the egg soon after the usual incubation period ends, and may start over.

Hatching

The mottled tan chick, with brown or black streaks on its head, hatches in 35 to 36 days on average. Time from the first crack in the egg, called a star fracture, to complete hatching is about five days. During this time, the chick's peeps from inside the shell are audible.

When a small chick falls from its perch, tern parents abandon it, even if it's in plain sight. If someone replaces the hatchling on or near its home perch, parents usually feed and warm it. If an older chick falls from the branch, parents often, but not always, find it and feed it on the ground. After a chick dies, parents given an

It took this hatching chick most of a day to wiggle free of its eggshell. Chicks are wet as they emerge. (DAVID DOW)

Chicks hatch covered in down. This chick's downy neck feathers are starting to dry while the rear of the bird is still in its shell. (DAVID DOW)

Chicks just a few hours old, such as this one being warmed by its parent, have well-developed feet and claws, crucial to clinging to exposed hatch sites.
(RICH DOWNS)

orphaned chick will sometimes feed and attend it, adopting the new chick as their own.

Once a chick is large enough to maintain its own body heat, both parents leave simultaneously to fish, leaving the chick alone.

Within hours of hatching, a downy chick can stand on its branch. If approached, the youngster rears up and spreads its wings in a defiant pose. This day-old chick defends its home. (DAVID DOW)

Only a couple of days old, this chick was left alone while its parents went fishing. (SUSAN SCOTT)

Fledging

On average, Honolulu chicks are 45 days old when they take their first flight. At this time, the youngsters still have visible down feathers. By flying at just over six weeks old, Honolulu chicks are remarkable. At Tern Island in French Frigate Shoals, chicks begin flying when

Chicks with downy feathers and stubby wings often flap vigorously while standing near their hatch sites. This Honolulu chick shared its branch with electrical wires. (ROBERT WEBER)

they are close to eight weeks old (average of 53 days), and at Ascension Island, chicks fledge in nine weeks (average 65–70 days).

When a White Tern chick starts taking short flights, it is 80 to 90 percent adult size, and its wings are not yet full-sized. In the 1990 book *Seabirds of Hawaii,* Craig Harrison writes, "Parent White Terns may literally push a chick off its perch for its first flight, and then fly alongside the fledgling." Parents may also tempt an

offspring with fish to encourage it to fly. The chick gains weight and wing length during the post-fledging period of life, remaining under parental care.

After it can fly, the chick stays in the vicinity of its hatch site, and the parents continue to feed it. Gradually, over a couple of weeks, the chick moves farther and farther from its birthplace and eventually becomes independent of its parents. The timing from fledging to independence varies from chick to chick and place to place. Some offspring permanently leave home three to four weeks after their first flights; others stick around the parents for up to 11 or so weeks.

(Above) Tan wing bars and patches of down are often visible on large chicks, such as this one. Watching a couple raise a chick to independence takes several months. Approximately two and a half to three months pass between the day the female lays an egg and the chick's first flight. Because it takes time for young terns to learn to catch their own fish, fledged chicks don't immediately leave home; rather, they stick around the family tree for various amounts of time, getting fish from their parents. (RICH DOWNS)

(Left) This medium-sized chick's feathers are mostly down. (SUSAN SCOTT)

The parents here seem to be discussing what to do with their wayward chick. The two fed the grounded youngster, and it lived to fly away. (SUSAN SCOTT)

Parental care

Both parents catch and carry fish and squid to their chick crosswise in their beaks, holding as many as 16 at a time. On Australia's Norfolk Island, parents, in one day, fed a 5- to 15-day-old chick 37 fish in 12 visits.

When a parent carrying fish lands near the chick, the chick approaches the parent with its head down and beak upturned. The chick pecks at, grasps, and tugs a single prey from the tip of its parent's bill. If a fish drops onto the branch, the parent picks it up and presents it again. If a fish falls to the ground, the parent abandons it.

As the chick grows, the size of fish offerings increases. If the fish is too big for the chick to swallow,

This parent is holding at least 8 fish. One Hawai'i seabird researcher recorded a White Tern carrying 16 fish and squid. (UNATTRIBUTED FRAMED PHOTO BOUGHT IN HALE'IWA DIVE SHOP, 2009.)

The multiple window ledges on a Tern Island building in the Papahānaumokuākea Marine National Monument are attractive to tern parents, but their angle and exposure may account for the low fledging rate there. During my work on Tern Island as a U.S. Fish and Wildlife volunteer, I often saw eggs roll off the ledges, with females re-laying time and again in the same spot. If the eggs hatched, chicks were usually able to hang on, but strong winds sometimes blew them to the ground. Workers patrolled daily to replace uninjured chicks on their ledges. (SUSAN SCOTT)

the parent supports the fish's tail as the chick proceeds to swallow it. When prey protrudes from the chick's mouth, the chick holds it there, digesting the head and body until it can swallow the rest of the fish.

Success rates

Honolulu's tern parents have an unusually high rate of success in raising a chick to fledging. A study in Kapi'olani Park between 1976 and 1984 revealed that of 120 eggs laid, 106 hatched, and of those hatchlings, 91 grew up to fly away. This is an 85 percent success rate. In another O'ahu study between 2001 and 2003, 65 of 88 eggs laid resulted in chicks fledging, a success rate of 74 percent. Some couples raised a second chick after the first fledged.

In Hawai'i's French Frigate Shoals Atoll, the success rate of White Terns raising a chick to independence was 30 percent. Ascension Island's success rate was 29 percent.

Threats

Weather

Strong winds sometimes blow eggs and chicks from their trees; chicks are also vulnerable to heavy rain, some dying of cold when their parents are fishing. Chicks left alone don't seek shelter during rainstorms, and parents don't appear to hurry back to protect their chick if the weather turns stormy.

Gusty tradewinds can dislodge precariously balanced eggs and newly hatched chicks when parents raise their bodies to change shifts.

Chicks must sit out storms alone. This chick survived a soaking rain. (SUSAN SCOTT)

During a blustery tradewind day, this chick braced itself between branches. (SUSAN SCOTT)

Predators

During the nineteenth century, wild bird feathers were in high demand for women's hats. Feather hunters killed White Terns and other seabirds by the hundreds of thousands, devastating colonies throughout the Northwest Hawaiian Islands. In 1903, President Theodore Roosevelt took action to protect seabirds from poachers, and in 1909 he created the Hawaiian Islands Bird Reservation. Since 1918, the White Tern, along with all other seabirds and shorebirds, has been fully protected under the Migratory Bird Treaty Act. In 1986, the State of Hawai'i listed the White Terns of O'ahu as a threatened species, likely because of its limited range and small population.

People in other parts of the world ate White Tern eggs and killed the birds for their snowy white feathers. As a result of such hunting, White Tern breeding colonies have been reduced mostly to offshore islands with cliffs and tall trees inaccessible to humans and other predators. Fortunately for O'ahu's residents and visitors, Honolulu has become a significant exception. White Terns in Honolulu don't seem bothered by either vehicles or people.

Today, in Hawai'i and most parts of the world, people are protective of White Terns. When Honolulu chicks fall from their branches, caring residents often successfully return them to the tree. Even if a chick is placed on a lower branch, parents usually continue to feed it.

Common bird predators such as cats, rats, and mongooses occasionally eat eggs and kill chicks and adults.

Four rodent species have been introduced to Hawai'i: the house mouse and three rats—brown, black, and Polynesian. All four rodents prey on birds, eating eggs, chicks, and adults, and all four can climb trees. Mice usually stay on the ground, and brown rats climb only a few yards. Polynesian and black rats, however, are expert climbers and regularly visit tall treetops. Pictured is a black rat, Rattus rattus, *also called a ship rat or roof rat.* (SUSAN SCOTT)

Even well-fed cats like these wild (feral) individuals at He'eia State Park instinctively hunt and kill birds. To help Hawai'i's birds, keep pet cats indoors and support island spay and neuter programs. (SUSAN SCOTT)

Hawai'i state officials introduced barn owls (Tyto alba) *from zoo stock in California and Texas in 1958 to eat rats and mice. The thriving nocturnal owls eat rodents, but also prey on native seabirds and shorebirds.* (TIM FELCE)

Although the city's terns may not be bothered by pedestrians, cars, lights, or noise, urban cats, rats, and mongooses are. The 24-hour-a-day activity in Honolulu, plus public and private rat-control efforts, may deter these predators from pursuing the birds.

Checking that rat-control stations have bait and keeping pet cats indoors help keep Honolulu's White Terns safe.

White Terns are generally serene, but if threatened, they fight back, either individually or in flocks. When a Peregrine Falcon arrived at Midway Atoll in 2015, White Terns gathered by the hundreds to mob it, screeching and pecking the falcon at high altitudes. Similar tern attacks occurred at Midway in 1978 against a Steller's

White Terns harass a perched peregrine falcon (lower left) at Midway Atoll, 2017.
(MARTHA BROWN)

On Tern Island, this White Tern defending a potential egg site on a concrete post flew at my raised camera and pecked my forehead. The injury was minor, but the message to back off was clear. (SUSAN SCOTT)

sea eagle, and in 2010 against a barn owl. White Terns have also been known to attack cats, dogs, and people. If humans, or other birds, get too close to a parent protecting an egg or chick, adults will sometimes swoop at the intruder, occasionally making contact.

On Oʻahu, barn owls, mynahs, bulbuls, and pigeons harass White Terns, causing some egg and chick losses. At Midway, mynahs occasionally evict a White Tern from its brooding perch and eat the egg. Introduced ants at Midway climb trees and attack and eat the webbing, eyes, and mouths of White Tern chicks.

Because White Terns depend on predatory fish driving small fish and squid to the surface, overfishing in Hawaiʻi can indirectly hurt the terns.

When Honolulu workers began a house demolition in 2007, a White Tern chick was discovered in the pictured kukui tree. Workers delayed the tree's removal but continued clearing. The commotion bothered neither parents nor offspring, and the chick lived to fledge. Tree felling or trimming should not occur until a chick is at least 45 days old after hatching, or roughly 80 days after an egg is laid. (KEITH SWINDLE/U.S. FISH AND WILDLIFE SERVICE)

Tree trimming

Honolulu hosts over 250,000 trees, most introduced from other parts of the world. One theory of why White Terns have adopted Honolulu as a breeding ground is the presence of trees with wide crowns, such as monkeypod, shower, kukui, banyan, and mahogany. The city's professional tree trimmers routinely trim these trees, sawing off dead branches for the safety of people below and clipping wayward boughs that threaten utility poles and wires.

If a tern family has set up housekeeping on a troublesome limb, however, cutting causes a dilemma. In addition to the fact that no tree trimmer wants to destroy a tern egg or kill a chick, it's also illegal. All of Hawai'i's seabirds, including the White Tern, are protected under both state and federal laws, and disturbing them can result in a fine.

Private and public agencies work with tree-trimming companies to identify egg and chick branches so workers can delay sawing until after a chick has fledged.

Conservation

Hui Manu-o-Kū

Even after Mayor Mufi Hanneman proclaimed the White Tern the official bird of Honolulu in 2007, a group of seabird conservationists noted that, several years later, little attention was being paid to the remarkable birds. As a result, in 2016 the group created the all-volunteer Hui Manu-o-Kū (*hui* means group in Hawaiian), a collaboration dedicated to raising public awareness of O'ahu's unique White Tern population. The *hui* includes representatives from Pacific Rim Conservation, U.S. Fish and Wildlife Service, Hawai'i Audubon Society, 'Iolani School, and devoted tern enthusiast Rich Downs, who has made White Tern research, photography, education, and management of the Hui his post-retirement career. Other interested citizens joined the group, the author included, each offering personal and professional skills toward the goals of increasing knowledge of and protecting these charming native seabirds.

Hui Manu-o-Kū's official mission statement is "to enhance awareness, appreciation, understanding and conservation of Manu-o-Kū."

Citizen Science

Because facts about White Terns are crucial in determining how to help them continue thriving in Honolulu, in 2016 Hui Manu-o-Kū began a Citizen Science program, which it continues to maintain and promote. Its volunteers, of all ages and abilities, participate in surveys of the city's terns, collecting and recording information for the Manu-o-Kū database. Rich Downs conducts training sessions and organizes watches.

Citizen scientists track the growing population of White Terns in the city by, among other things, reporting new eggs, recording egg incubation times, documenting behaviors, and noting interactions with predators. Because White Terns are so photogenic, both professional and amateur photographers take extraordinary pictures, some of which are added to the Hui's online database. Pictures of tern parents feeding their chicks help researchers learn more about the marine species that inhabit Hawai'i's offshore waters.

You can see nest maps, egg maps, and a complete O'ahu breeding range map, and learn more about White Terns, the Hui, and the Citizen Science program at www.whiteterns.org.

In its beak tips, this parent holds a Freckled Driftfish (Psenes cyanophrys), *a tropical and subtropical species found worldwide. Because the fish swims near, or under, floating seaweed or other objects afloat, people rarely see it. Biologist Sarah Donahue identified the fish.* (RICH DOWNS)

Helping arborists

Although tree trimming can be a threat to White Terns, it also helps them by removing dead branches that could fall on eggs or chicks. Cutting away live branches in dense parts of the trees creates open spaces for birds to fly in and out. In addition, trees grow scars around cut sites, forming cuplike hollows ideal for holding eggs. It's these circles of wood that, when upright, attract tern parents looking for a place to balance an egg and raise a chick. The branch scars also give chicks ridges to grip with their sharp toenails.

Tree-trimming scars like this one help White Tern parents balance their egg. (RICH DOWNS)

Besides warning workers to trim with caution while Manu-o-Kū are actively breeding, blue trunk tape displays the phone number for the White Tern Hotline, sponsored by the Hui, *and the URL for their webpage, which offers guidelines about what to do if a chick falls from the tree.* (SUSAN SCOTT)

Sidewalk spots like these on a King Street curb are a clue that White Terns are breeding above. White droppings are typical of seabirds. (SUSAN SCOTT)

After terns have already laid an egg or hatched a chick, however, trimming that tree should be delayed. A tree must remain untrimmed for at least 80 days after a female lays an egg. Members of Hui Manu-o-Kū and Hawaiʻi's Department of Land and Natural Resources work together to tie sky-blue "tern tape" around tree trunks to alert trimmers and arborists to the presence of breeding terns.

Before scheduling tree trimming, view the active tern nest map at https://www.whiteterns.org/active-manu-o-k363-nest-map.html for known breeding activity in the tree. To check a tree for breeding terns, look below the branches for the all-white droppings typical of fish-eating seabirds.

Helping injured or abandoned birds

If you find a live chick on the ground and know where it came from, try to put it back. If you can't find its home, or the tree is too tall for safe climbing, call the Manu-O-Kū hotline at 808-379-7555 or the U.S. Fish and Wildlife Service Office of Law Enforcement at 808-861-8525. Go to www.whiteterns.org for more information.

Do not attempt to feed or give water to a grounded bird. If the bird is dead, check the leg for a band and report it to one of the numbers above.

A parent warms its chick in a Honolulu tree. The late Hawai'i naturalist George C. Munro wrote in 1941, "It seems possible that the species is slowly working back to the main group [of the Hawaiian Islands]. We should do everything to encourage and help it in this." (CLAUDE MATSUMOTO)

Manu-o-Kū festival

In 2016, the Conservation Council for Hawaiʻi's executive director, Marjorie Ziegler, organized the first White Tern Festival to increase awareness of Honolulu's White Terns. The gathering gave birth to Hui Manu-o-Kū, and in 2017 the Conservation Council, the Hui, individual citizens, and both private and public organizations put on a second Manu-o-Kū Festival to celebrate Honolulu's official bird.

The family-friendly celebration is now an annual event featuring a day of entertainment and education for adults and children alike. Please visit the Hui Manu-o-Kū website for the date, location, and time of this annual springtime celebration of Oʻahu's beautiful terns.

The Future

The following are ways residents and visitors can help Honolulu's White Terns continue to thrive:

- Become a White Tern Citizen Scientist. See Hui Manu-o-Kū's website tab "Citizen Science."
- Learn about the city's White Terns and spread the word. An enjoyable way to discover where the birds are, and how they're currently doing, is to join Rich Downs on his informative tern walks. Dates, times, and locations are posted on the Hui Manu-o-Kū website, under the tab "Manu-o-Kū News," and on the Hawaiʻi Audubon Society's website under the tab "Events/Get Outside."

Some O'ahu residents go out of their way to help White Terns. In a tree on Hawai'i State Capitol grounds, a tree trimmer fastened this wood rail to help parent terns balance their egg and their subsequent chick on the unstable site. (SUSAN SCOTT)

- Help arborists and homeowners learn tree-trimming guidelines, found on the Hui Manu-o-Kū website under the tab "Get to Know Manu-o-Kū: Tree Trimming Information." Get help with strategies under the "Contact" tab.
- Engage schoolchildren in the story of Honolulu's White Tern.
- Take pictures of White Terns, noting exactly where they have laid an egg or hatched a chick, and send the photos and information to Hui Manu-o-Kū at huimanuoku@gmail.com.
- Donate to and/or volunteer with conservation organizations that fund White Tern research.

References

Ainley, David G., et al. 2015. "Seabird flight behavior and height in response to altered wind strength and direction." *Marine Ornithology* 43: 25–36.

Amerson, A. Binion, Jr. 1971. "The Natural History of French Frigate Shoals, Northwestern Hawaiian Islands." Atoll Research Bulletin, No. 150 issued by the Smithsonian Institution with the assistance of the Bureau of Sport Fisheries and Wildlife U.S. Department of the Interior, Washington DC.

Ashmole, Philip N. 1968. "Breeding and Molt in the White Tern (*Gygis alba*) on Christmas Island, Pacific Ocean." *Condor* 70: 35–55.

Aymar, Gordon C. 1938. *Bird Flight.* Garden City, NY: Garden City Publishing Company. P. 234.

Baker, Allan J., et al. 2007. "Phylogenetic relationships and divergence times of *Charadriiformes genera:* Multigene evidence for the Cretaceous origin of at least 14 clades of shorebirds." *Biology Letters* 3 (2): 205–209.

Burger, J., et al. 2001. "Metals in feathers of Sooty Tern, White Tern, Gray-backed Tern, and Brown Noddy from islands in the North Pacific." *Environmental Monitoring and Assessment* 71: 71–89.

Byrd, Vernon G., and Thomas C. Telfer. 1980. "Barn owls prey on birds in Hawaii." *'Elepaio* 41 (5): 35–36.

Carlile, Nicolas, and David Priddel. 2015. "Establishment and growth of the White Tern *Gygis alba* population of Lord Howe Island, Australia." *Marine Ornithology* 43: 113–118.

Croxall, J. P., et al. 1984. "Status and Conservation of the World's Seabirds." International Council for Bird Preservation (ICBP) Technical Publication No. 2, Cambridge, England.

Dorward, Douglas F. 1961. "Comparative breeding biology of some seabirds of Ascension Island, with special reference to the species of Sula and the Fairy Tern." Doctor of Philosophy Thesis, University of Oxford, Edward Grey Institute, Botanic Garden, Oxford, England.

Dorward, Douglas F. 1963. "The Fairy Tern *Gygis alba* on Ascension Island." *Ibis* 103 (3): 365–378. (Abstract. No full text available.)

Downs, Richard, and Keith Swindle. 2017. "State Art Museum Renovation Schedule Yields to Pair of Nesting White Terns." *'Elepaio* 77 (3): 20.

Fefer, S. I., et al. 1984. "Synopsis of results of recent seabird research conducted in the Northwestern Hawaiian Islands." In: R. W. Grigg and K. Y. Tanoue (eds). Proceedings of the second symposium on resource investigations in the Northwestern Hawaiian Islands, Vol 1, Sea Grant Miscellaneous Report UNIHI SEAGRANT – MR-84–01. Honolulu. Pp. 9–76.

Harrison, Craig S. 1990. *Seabirds of Hawaii, Natural History and Conservation*. Ithaca, NY: Cornell University Press.

Harrison, Craig S., et al. 1983. "Hawaiian Seabird Feeding Ecology." *Wildlife Monographs,* a Publication of the Wildlife Society, (85): 3–71.

"Hawaii's Comprehensive Wildlife Conservation Strategy." 2005. "Seabirds, Manu-o-Kū or White (Fairy) Tern, *Gygis alba*." Hawai'i Department of Land and Natural Resources, Honolulu. P. 358.

Hui Manu-o-Kū. 2018. https://www.whiteterns.org.

Jeong, Min-Su, et al. 2014. "The First Record of the White Tern (*Gygis alba*) in South Korea." *Korean Journal of Parasitology* 21 (2): 69–73.

Martinez-Gomez, Juan E., and Noemi Matias-Ferrer. 2013. "First Breeding Record of the White Tern, *Gygis alba,* in Mexico." *Wilson Journal of Ornithology* 12 (4): 844–846.

Miles, Dorothy H. 1986. "White Terns Breeding on Oahu, Hawaii." *'Elepaio* 46 (16): 171–175.

Morgan, Lydi, 2007. "Manu-o-Kū named the official bird of Honolulu." *'Elepaio* 67 (4): 25–27.

Munro, George C. 1941. "Birds of Hawaii and Adventures in Bird Study: The White or Love Tern." *'Elepaio* 10: 1–5.

Murakami, Linda D. 1977. "Seabirds in the City: Breeding of the White Tern in Lower Makiki." *'Elepaio* 38 (6): 63–65.

Niethammer, Kenneth R. 1998. "White Tern (*Gygis alba*)." *The Birds of North America Online* (A. Poole, ed.) Ithaca, NY: Cornell Lab of Ornithology. Retrieved from *The Birds of North America Online:* http://bna.birds.cornell.edu/bna/species/371, 2009.

Ord, Michael. 1961. "White Tern at Koko Head, Oahu." *'Elepaio* 22 (3): 17–18.

Pukui, Mary Kawena, and Samuel H. Elbert. 1986. *Hawaiian Dictionary: Hawaiian–English and English–Hawaiian, Revised and Enlarged Edition*. Honolulu: University of Hawai'i Press.

Rauzon, Mark J., and Karl W. Kenyon. 1984. "White Tern nest sites in altered habitat." *'Elepaio* 44 (8): 79–80.

Spear, Larry B., and David G. Ainley. 1997. "Flight behavior of seabirds in relation to wind direction and wing morphology." *Ibis* 139: 221–233.

Stanton, Joseph. 2006. *A Field Guide to the Wildlife of Suburban O'ahu*. St. Louis: Time Being Books.

Surman, C. A., and R. D. Wooler. 2003. "Comparative foraging ecology of five sympatric terns at a sub-tropical island in the eastern Indian Ocean." *Journal of Zoology* 259 (3): 219–230.

Thompson, Nainoa. (Undated) "Hawaiian Voyaging Traditions: On Wayfinding." http://archive.hokulea.com/ike/hookele/on_wayfinding.html.

Tinker, Spencer. 1978. *Fishes of Hawaii: A Handbook of the Marine Fishes of Hawaii and the Central Pacific Ocean*. Honolulu: Hawaiian Service, Inc.

Tomich, P. Q. 1971. "Notes on Barn Owl in Hawaii." *'Elepaio* 23: 16–17.

U.S. Fish and Wildlife Service. 2005. "Regional Seabird Conservation Plan, Pacific Region." U.S. Fish and Wildlife Service, Migratory Birds and Habitat Programs, Pacific Region, Portland, Oregon.

U.S. Fish and Wildlife Service. 2018. "Midway Seabird Protection Project: Draft Environmental Assessment: Sand Island, Midway Atoll, Papahānaumokuākea Marine National Monument." https://www.fws.gov/uploadedFiles/Region_1/NWRS/Zone_1/Midway_Atoll/Sections/What_We_Do/Resouce_Management/Midway_Seabird_EA_Public_Draft.pdf.

VanderWerf, Eric A. 2003. "Distribution, Abundance, and Breeding Biology of White Terns on Oahu, Hawaii." *Wilson Bulletin* 115 (3): 258–262.

VanderWerf, Eric A., and Richard E. Downs. 2018. "Current distribution, abundance, and breeding biology of White Terns (*Gygis alba*) on Oahu, Hawaii." *Wilson Journal of Ornithology* 130 (1): 297–304.

Wegmann, Alexander, et al. 2008. "Pacific rat *Rattus exulans* eradication on Dekehtik Island, Federated States of Micronesia, Pacific Ocean." *Conservation Evidence* 5: 23–27.

Wegmann, Alexander, et al. 2008. "Ship rat *Rattus rattus* eradication on Pein Mal Island, Federated States of Micronesia." *Conservation Evidence* 5: 28–32.

Wegmann, Alexander, et al. 2014. "Rats to Palm Trees: Baiting the Canopy during the Palmyra Atoll Eradication Project." Proceedings of the 26th Vertebrate Pest Conference (R. M. Timm and

J. M. O'Brien, eds.). Published at University of California, Davis. 73–77.
Winship A. J., et al. 2016. "Marine Biogeographic Assessment of the Main Hawaiian Islands, Chapter 7: Seabirds." *Bureau of Ocean Energy Management and National Oceanic and Atmospheric Administration. NOAA Technical Memorandum NOS NCCOS 214:* 283–319.
Yeung, Norine W. 2010. "Systematics, phylogeography, and feeding ecology of the white tern (*Gygis alba*)." Thesis for the degree of Doctor of Philosophy (University of Hawai'i at Mānoa), Zoology (Ecology, Evolution, and Conservation Biology); no. 5454.
Yeung, Norine W., et al. 2009. "Testing subspecies hypothesis with molecular markers and morphometrics in the Pacific white tern complex." *Biological Journal of the Linnean Society* 98: 586–595.
Young, H. S., et al. 2010. "Niche partitioning among and within sympatric tropical seabirds revealed by stable isotope analysis." *Marine Ecology Progress Series* 416: 285–294.

About the Author

A former registered nurse, **SUSAN SCOTT** earned a bachelor's degree in biology from the University of Hawai'i in 1985 and is a graduate of the university's Marine Option Program, where she specialized in marine science journalism. Since 1987, Susan has written a weekly column called "Ocean Watch" for the *Honolulu Star-Advertiser* and has worked as a volunteer for the U.S. Fish and Wildlife Service since 1989. This is her ninth book about nature in Hawai'i.

For questions, comments and/or corrections contact Susan via the email link at www.susanscott.net.